Our Only World

Poetry for Planet Earth

Poets Unite Worldwide

Fabrizio Frosini

Poems by

Alexandro Acevedo Johns, Chile

Ellias Aghili Dehnavi, Iran

Saadat Tahir Ali, Pakistan/KSA

Anna Banasiak, Poland

Tom Billsborough, UK

Daniel J. Brick, USA

Mayjorey Buendia, The Philippines

Fabrizio Frosini, Italy

Alem Hailu G/Kristos, Ethiopia

Nosheen Irfan, Pakistan

Afrooz Jafarinoor, Iran

Farzad Jahanbani, Iran

Srijana KC Rayamajhi, Nepal

Joji Varghese Kuncheria, India/Oman

Tapera Makadho, Zimbabwe

Mallika Menon, India

Barry Middleton, USA

Leloudia Migdali, Greece

Bharati Nayak, India

Madhumita Bhattacharjee Nayyar, India

Valsa George Nedumthallil, India

Mohammed Asim Nehal, India

Anil Kumar Panda, India

Marcondes Pereira Da Silva De Mesquita, Brazil

Sarah Louise Persson, UK

Sajee Rayaroth, Australia

Marianne Larsen Reninger, Denmark/USA

Kirti Sharma, India

Anzelyne Shideshe, Kenya/Germany

Pamela Sinicrope, USA

Kim Solem, USA

Udaya R. Tennakoon, Sri Lanka/Switzerland

Tze Min Ition Tsai, Taiwan

Savita Tyagi, India/USA

Hans Van Rostenberghe, Belgium/Malaysia

Joey Vlahakis, USA

Our Only World
Poetry for Planet Earth

Poets Unite Worldwide
Fabrizio Frosini

Copyright © 2017, 2018 Fabrizio Frosini
1st paperback edition 2018
Independently published by Fabrizio Frosini

Editorial project by Fabrizio Frosini
Anthology of Poetry – Contributors:
Fabrizio Frosini, Pamela Sinicrope, Daniel J. Brick, Alexandro Acevedo Johns,
Ellias Aghili Dehnavi, Saadat Tahir Ali, Anna Banasiak, Tom Billsborough,
Mayjorey Buendia, Alem Hailu G/Kristos, Nosheen Irfan, Afrooz Jafarinoor,
Farzad Jahanbani, Srijana KC Rayamajhi, Joji Varghese Kuncheria, Tapera
Makadho, Mallika Menon, Barry Middleton, Leloudia Migdali, Bharati Nayak,
Madhumita Bhattacharjee Nayyar, Valsa George Nedumthallil, Mohammed Asim
Nehal, Anil Kumar Panda, Marcondes Pereira Da Silva De Mesquita, Sarah Louise
Persson, Sajee Rayaroth, Marianne Larsen Reninger, Kirti Sharma, Anzelyne
Shideshe, Kim Solem, Udaya R. Tennakoon, Tze Min Ition Tsai, Savita Tyagi, Hans
Van Rostenberghe, Joey Vlahakis

Cover: image from Pixabay.com, used under Creative Commons CC0

ISBN: 9781980553229

«No, I would not agree that it [*carbon dioxide*]
is a primary contributor to the global warming that we see.»
[March 2017]

«Paris [*the Paris climate agreement*] is something we need to look at closely.
It's something we need to exit in my opinion,»
[April 2017]

Scott Pruitt
Trump's advisor and head of Environmental Protection Agency (EPA)
(about carbon dioxide level in atmosphere and the Paris climate agreement)

TABLE OF CONTENTS

«On the fifth day
the scientists who studied the rivers
were forbidden to speak
or to study the rivers.
The scientists who studied the air
were told not to speak of the air,
and the ones who worked for the farmers
were silenced,»

Jane Hirshfield, "On the fifth day"
[*poem read from the stage at the March for Science on April 22, 2017*]

ACKNOWLEDGEMENTS

Thirty-six poets, from twenty-two countries, have taken part in this compilation. They belong in the free Association 'Poets Unite Worldwide', and have answered my call both for their love for Poetry and for their deep feelings towards environmental issues. My gratitude to each of them.

Indeed, the Earth is the only world we have, and the poems of this collection, filled with empathy and sensitivity, thoroughly show our deep concern for the future both of our planet and of humankind — a future that the executive orders coming from the White House make even less comforting.

Although a product of the collaborative effort of many poets, this collection has become a book thanks to the work of my talented coeditors: Pamela Sinicrope and Margaret O'Driscoll. To them, my special thanks.

Ad maiora!

(Fabrizio Frosini, Italy)

Here lies the body of this world,
Whose soul alas to hell is hurled.

Henry David Thoreau, "Epitaph on the World"

INTRODUCTION

Climate change, wildlife collapse, environmental destruction: just a trivial science-fiction movie? Unfortunately, this is already part of our present, and to an even greater extent will be our generation's legacy to our grandchildren. Land, water and air pollution is so widespread that nowhere *"a truly virgin environment"* still exists: man-made waste has contaminated the entire globe, and together with the increased levels of carbon dioxide in the atmosphere and in the oceans, they have become markers of a new geological epoch: ***Anthropocene.***

Carbon dioxide in the atmosphere.

According to a detailed global analysis from the World Meteorological Organization (WMO), climate change is reaching disturbing new levels across the Earth, showing "extreme and unusual" conditions. As David Carlson, World Climate Research Programme Director at the WMO said: *"Even without a strong El Niño in 2017, we are seeing other remarkable changes across the planet that are challenging the limits of our understanding of the climate system."* Thus, as stated by Phil Williamson, from the University of East Anglia: *"Human-driven climate change is now an empirically verifiable fact, combining year-to-year variability with the consequences of our release of extra greenhouse gases. Those who dispute that link are not sceptics, but anti-science deniers."*
Unfortunately, the prominent denier of global warming, today, is the Trump administration: the present head of the US Environmental Protection Agency (EPA), Scott Pruitt, publicly denies that CO2 is a primary contributor to global warming —this way the head of EPA negates EPA policy and 'overwhelmingly clear' evidence on climate change. Furthermore, President Trump's budget blueprint for the coming fiscal year would slash the Environmental Protection Agency by 31%! Is this just 'politics'? Or we want to call it blindness, madness..? For many, Mr. Trump is that kind of person who views not only America, but the entire world as a showcase for their own ego. Regrettably, his closest allies and advisers seem

to be of the same kind…

On March 13, 2017, thirty prominent climate scientists sent a letter to Scott Pruitt, refuting his false statement:"*Just as there is no escaping gravity when one steps off a cliff, there is no escaping the warming that follows when we add extra carbon dioxide and other greenhouse gases to the atmosphere,*" the scientists wrote [see bibliography]. The group included Nobel laureate chemist Mario Molina of the University of California, San Diego, and members of the National Academy of Sciences (US).

As Marianne Lavelle wrote: "*Under President Barack Obama, the United States pledged to reduce greenhouse gas emissions 26-28 percent from 2005 levels by 2025. That means emissions must be cut about 1.7 billion metric tons, according to figures from the Environmental Protection Agency's latest greenhouse gas inventory. The nation is a third of the way to that target, but the rest was to be achieved via an array of regulations, especially the Clean Power Plan, that are now targeted for elimination by President Donald Trump. Not only was the goal dependent on those rules, it would have also required even more rigorous policies from Obama's successor because reductions from those rules would not have been enough, numerous studies have found.*" On the contrary: "*President Donald Trump's planned climate change policies could lead to an extra half a billion tons of greenhouse gases in the atmosphere by 2025, according to a new analysis. That is equal to the annual electricity emissions of 60 percent of U.S. homes.*" [Marianne Lavelle, 'Inside Climate News', March and April 2017].

And we know that should the United States exit the Paris climate agreement, other countries will not retain the commitments made in Paris.

About global warming, we need to consider that the oceans are absorbing about 93% of the increase in the Earth's heat. Compared with the 1961-1990 reference period, 2016 was 0.83 degrees C warmer than the average (WMO's State of the Global Climate 2016 report). Yet, in the Arctic, temperatures were about 3 degrees C above the 1961-1990 average; and in Svalbard, in the Arctic circle, the yearly average was 6.5 degrees above the long-term mark.

Last January, a panel of NASA scientists provided data showing that sea levels worldwide rose an average of nearly 3 inches (8 cm) since 1992, as a result of warming waters and melting ice. This means that the predicted sea level rise by the end of the century, will be likely close to 3 feet (almost one meter) and maybe higher. To catch the whole meaning, we need to consider that, nowadays, more than 150 million people, mostly in Asia, live within those 3 feet of the sea.

Carbon dioxide in the oceans.

The state of coral reefs is a telling sign of the health of the seas. The increased levels of carbon dioxide in the oceans have other awful effects on the environment: last year, hundreds of miles of Australia's Great Barrier Reef most pristine northern sector, were found to be dead, killed by the warming of water. Furthermore, an even greater extension of reef [a 900 miles (1,500 km) stretch of the reef, according to scientists] shows 'bleaching' —bleaching happens when high water temperatures cause corals to expel the algae they depend upon and give them color; if normal conditions return, the corals can recover, but it can take decades; yet, if the stress continues the corals die. The Australian government confirmed (March 2017) that widespread coral bleaching is happening on the Great Barrier Reef for the fourth time.

Yet, the heating up, acidification, and the loss of oxygen are not the only negative effects of climate change on the oceans; many other events can alter the balance of the marine environment. Among them: toxic algae blooms may become more common, and when billions and billions of toxic cells come together, they can easily poison the environment, kill marine life, and economically harm entire coastal communities.

Plastics.

Plastics is all around us.. Each year, each of us, on average, uses more than 200 plastic bags, which can take between 100 and 400 years to degrade. Moreover, plastics usage is increasing, with an annual production exceeding 300 million tonnes, and over a third of that is used to make items of packaging, which are then rapidly discarded. As an international study has revealed, since WW2 years we have made enough plastic to coat the Earth entirely in cling film!

Although plastics bring many societal benefits and offer technological advances, their accumulation in landfills and in natural habitats means awful problems for wildlife, resulting from plastics ingestion, the leaching of chemicals from plastic products and the potential for plastics to transfer chemicals to wildlife and humans. Particularly of great concern for their profound environmental effects are microplastics: fragments as small as 1.6

μm have been identified in many marine habitats —even in the very depths of the oceans— where they can be easily ingested by small marine invertebrates. A recent video footage by a scientist, massively shared through media and social networks, shows fluorescent pieces of plastic being sucked in by plankton and moving into their guts [*link to 'video' in bibliography*]. This way, plastic has the potential to transfer toxic substances to the food chain —most plasticizers appear to act by interfering with hormone function, and molluscs, crustaceans and amphibians appear to be particularly sensitive to such compounds. And eating contaminated seafood, we ingest our dose of microplastics too.

A World Economic Forum study found that there are currently at least 150 million tons of trash in the ocean. Besides, an United Nations scientific panel has estimated that about 8 million tons of plastic trash are dumped into the ocean every year, and that there are 46,000 pieces of waste plastic per square mile of sea: that means that in the oceans there are as many as 51 trillion particles of microplastics —500 times as many stars estimated to be in our galaxy. As Prof. Jan Zalasiewicz, of Leicester University, points out: "*A vast proportion of* [fish] *now have plastic in them. They think it is food and eat it, just as seabirds feed plastic to their chicks. Then some of it is released as excrement and ends up sinking on to the seabed. The planet is slowly being covered in plastic.*"

A huge number of animals, especially in the oceans, ranging from seabirds to turtles become entangled in plastic and drown or choke to death. Indeed, images of seals, turtles and seabirds trapped in plastic rings, ropes and sheeting are more and more common to see, and many tropical "*paradises*" show beaches littered with the bodies of chicks whose bellies were filled with plastics. But also large animals remain victims of our garbage: two months ago, a report from Norway described the sad death of a 20 foot long (c. 6 meters) adult Cuvier's beaked whale. The poor animal repeatedly beached and had to be euthanized; post-mortem examination revealed a lot of plastic bags and other items (man-made waste) in its stomach —according to researchers, the animal may have thought the bags were squid it could eat. As a Norwegian scientist said: "*It wasn't like it was in just part of the stomach,* [the waste] *filled up the whole space.*" [see bibliography for the link to the article].

Desertification.

Increasingly hot conditions in the Sahara desert resulting from man-made global greenhouse gas emissions, are producing both an increase in desertification and a strengthening of the violent Sahelian storms (some of the most explosive storms in the world). As Professor Christopher Taylor, meteorologist at CEH (Centre for Ecology & Hydrology, UK), says: "*Global warming is expected to produce more intense storms, but we were shocked to see the speed of the changes taking place in this region of Africa.*" —and the Sahel, a vast semiarid region south of the Sahara, is home to some of the most vulnerable populations on the planet.

From the UN web page on desertification:

— Due to drought and desertification, each year 12 million hectares are lost (that is 23 hectares/minute);

— Land degradation affects 1,5 billion people globally;

— 74% of the poor (42% of the very and 32% of the moderately poor) are directly affected by land degradation globally;

— Some two billion people depend on ecosystems in dry-land areas, 90% of whom live in developing countries.

A downward spiral is created in many underdeveloped countries where overpopulation causes pressure to exploit dry lands for farming. These marginally productive regions are overgrazed, the land is exhausted and groundwater is overdrafted. When rural land becomes unable to support the local population the result is mass migrations to urban areas. The increased frequency and severity of droughts resulting from projected climate change is likely to further exacerbate desertification.

In such circumstances, inequality is destined to explode. So as wars for water and land. In many countries, likely, human society will be deeply shaken.. upset..

The rate at which we're currently causing climate change is alarmingly fast —much faster than in most natural climate change events—, and the scientific observations highlight how temperature increases can produce sudden and drastic environmental impacts. Indeed, each and every day we should keep well in mind that since 1970, we have wiped out more than half of the Earth's wildlife populations, and that global warming is impacting every species on the planet, even down to the genetic level.

The result of all our actions on the environment is that we've pushed extinction rates of flora and fauna far above the long-term average: as researchers say, the Earth is now on course for a sixth mass extinction which would likely see 75% of species extinct in the next few centuries, if current trends continue. And our own existence, as superior animals — civilized humans— is at risk.

(*Fabrizio Frosini*)

Bibliography:

– *Barnes D.K.A. et al.*, "*Accumulation and fragmentation of plastic debris in global environments*", The Royal Society 2009
– *Cook K.H. et al.*, "*Detection and analysis of an amplified warming of the Sahara Desert*", J. Clim. 2015
– *Cozar A. et al.*, "*Plastic debris in the open ocean*", Proc. Natl. Acad. Sci. USA 2014
– *Hurley R. et al.*, "*Microplastics contamination of riverbeds*", Nature Geoscience 2018
– *Intergovernmental Panel on Climate Change (IPCC)*, "*Climate Change 2014 Synthesis Report - Summary for Policymakers*" :
https://www.ipcc.ch/pdf/assessment-report/ar5/syr/AR5_SYR_FINAL_SPM.pdf
– *Molina M. et al.*, "*Letter to S.Pruitt*" :
https://www.documentcloud.org/documents/3516598-Letter-to-Administrator-Pruitt.html
– *Park J. et al.*, "*Anthropogenic Mediterranean warming essential driver for present and future Sahel rainfall*", Nature 2016
– *PhysOrg article on the stranded whale in Norway:*
https://phys.org/news/2017-02-norway-zoologists-plastic-bags-stranded.html
– *Taylor C.M. et al.*, "*Frequency of extreme Sahelian storms tripled since 1982 in satellite observations*", Nature 2017
– *Thompson R.C. et al.*, "*Plastics, the environment and human health: current consensus and future trends*", The Royal Society 2009
– *UN web page on desertification:*
http://www.un.org/en/events/desertificationday/background.shtml
– *Video (in BBC article)* : *http://www.bbc.com/news/science-environment-39217985*
– *Zalasiewicz J. et al.*, "*The geological cycle of plastics and their use as a stratigraphic indicator of the Anthropocene*", ScienceDirect 2016

Small, rocky planet
Orbiting a mundane star
A galactic speck
Trivial piece of flotsam
Center of our universe.

Kelly Kurt

THE POEMS

Alexandro Acevedo Johns, *'Cry for Earth's Agony'*

Ellias Aghili Dehnavi, *'Let's respect Life'*

Saadat Tahir Ali, *'Pearly Mother'*

Anna Banasiak, *'Armageddon'*

Tom Billsborough, *'Global Warning'*

Daniel J. Brick, *'A Demi-God Speaks For The Earth'*

Mayjorey Buendia, *'Gaia'*

Fabrizio Frosini, *'Anthropocene Epoch'*

Alem Hailu G/Kristos, *'A lake's obituary'*

Nosheen Irfan, *'Earth is Crying'*

Afrooz Jafarinoor, *'The Last Days of Planet Earth'*

Farzad Jahanbani, *'This Woman Is'*

Srijana KC Rayamajhi, *'Wake up!'*

Joji Varghese Kuncheria, *'Save the Earth'*

Tapera Makadho, *'Epitaph For Mother Earth'*

Mallika Menon, *'Woes Of Mother Earth'*

Barry Middleton, *'Eden's Rose'*

Leloudia Migdali, *'Time to protect this Earth'*

Bharati Nayak, *'The Earth'*

Madhumita Bhattacharjee Nayyar, *'For the Love of Mother Earth'*

Valsa George Nedumthallil, *'Planet Earth'*

Mohammed Asim Nehal, *'Your Generosity and Our desires'*

Anil Kumar Panda, *'The Simmering Earth'*

Marcondes Pereira Da Silva De Mesquita, *'Unbearable future'*

Sarah Louise Persson, *'Earth's Survival'*

Sajee Rayaroth, *'Timeless In Time'*

Marianne Larsen Reninger, *'Work for the Night is Coming'*

Kirti Sharma, *'The Tree'*

Anzelyne Shideshe, *'Father Nature'*

Pamela Sinicrope, *'To the Polar Bear'*

Kim Solem, *'I Believe'*

Udaya R. Tennakoon, *'Wonderful creatures!'*

Tze Min Ition Tsai, *'The sunset even feel cold'*

Savita Tyagi, *'The Dragon's Mouth'*

Hans Van Rostenberghe, *'Fever!'*

Joey Vlahakis, *'Species'*

seasons change
but this spring is no more spring...
whale's bones with plastics

Fabrizio Frosini, Haiku
(in 'A Season for Everyone – Haiku & Tanka Poetry')

THE POEMS

Alexandro Acevedo Johns

Cry for Earth's Agony

The climate change threatens the World,
But we will be guilty of its crumbling.

The gifts of Earth that Virgil sang and Chinese and Indian poets
No longer can stand our delirium.
The dialogue of leaves in the forests, the naked run of the waters
And the immaculate whistling in the wind,
Sung by Neruda and Walt Whitman,
Are turned off in the ruckus of our pollution.

We have the Earth in our hands, like a woman dressed in garlands
That we tear off with violence, and we abandoned in the desert,
Without, see that we are alive thanks to her surrender.

It is not possible that our survival become a curse for Earth,
It is not possible that we are tormenting the spring
And let's be killers even of the bees.
We are simmering the air with our throwaway inventions.
We build pyramids of garbage

For the astonishment of future generations.
We have invoked a climate that is a threat to its masters
As if we were magicians with our wit in madness.

Because Humanity exists, is everything allowed to us?
I apologize for my words:
If there is someone created to God's likeness, it is Earth.
We are looking for in the infinite sky planets similar to it
Or at least with a drop of life, but is useless,
We only find dust of fire and desolate craters
Over stellar worlds of silence.

We don't cry out about mirages written over an imaginary wall,
The evidences are nailed inside our reality:
We are responsible for the suffering of Earth,
Its agony is the blinding reflection of our lifestyle.

Ellias Aghili Dehnavi

Let's respect Life

We are the assiduous race
In dimming our planet's face,
An aspersion casted on us,
we even may kill the planet Mars!

Our earth is filled with fumes
Hard for a blossom to bloom
In such a dark air
Where just a few show care.

In here it's so warm
That lands to seas transform
Our kids deride us for our deeds
Poor earth! His heart bleeds...

Poor polar bears
Please in your prayers
Don't forget their kids
Sink in ocean for our deeds.

For there'll be no poles

No beauty man extols

Earth is filling up with water

As we watch it getting hot and hotter.

For the future generations please

Let's save oceans, air and trees

Let's respect Life

With Earth let's not strife.

Saadat Tahir Ali

Pearly Mother

Paper I read, tossed to the side.

Can that I guzzled, falls out of my ride.

Shoppers I fling, lie in a heap.

Such piling makes our Earth Mother weep.

It's been there nearly a week

As we pass by with only a shifty peek.

Must we wade in knee deep rising muck,

Is this our life? We casually decree, it sucks.

This? It's done! It's fit to burn.

Oh That! Just sweep it under the fern.

Rubbish we make, our piles of rot.

On mother's face a growing blot.

I wonder where our children will go.

Enclosed by trash, how can they grow?

Disgusting and dirty, we all sing

Don't care to sort or do our thing.

We throw away trash after the binge!

Then recklessly watch it float or cling.

Walk past, sniffle and exult in turn.

Next-door bum! He'll never learn.
The truth is sad, harsh and bitter.
We resign to life packed with litter!
Care not enough to hear her sighing.
Our pearly Mother is surely dying.

Anna Banasiak

Armageddon

the colors of the world are changing

dark images hit as waves

of the Armageddon

life goes back

extinguished

disappears in the flood of pollution

I watch the sea level rise

the ice cap is melting

no one can breathe

under the burnt paradise

the earth is suffocating

in the lonely planet

in silence

everything

will vanish

in the face

of change

Tom Billsborough

Global Warning

Can we deselect stupidity?

Coral blanches in the warming seas,

The Arctic cap recedes,

Indifference breeds

An expanding waist of greed.

This is no Greek tragedy.

This is not great Sophocles,

Not Oedipus destroying Mother Earth.

That was accidental. This far worse

As the white faced clown

Is chosen from the chorus,

Merely an Apprentice,

Plucked for the task

By the call of Avarice.

Daniel J. Brick

A Demi-God Speaks For The Earth

"Who speaks for the Earth?"
(Carl Sagan)

Let's divide the world between us
and vie to see who loves their portion
most sincerely. But how will we
judge our passions, or rather the depth
of our passions? Is there a calculation
in mathematics to determine degrees
of love? Unlikely? What if I place my right
hand over my heart, and my left hand
palm down on the Earth, will I be able
to measure Love in terms of a rhythm
common to heart and ground?
Perhaps we should seek a simpler test:
All four hands palming the Earth,
we will look deep into each other's eyes,
and find in their depths the true
dimensions of love. And I'm certain
we will discover in each other equal love of Earth.

Let's sit for awhile on the woodchips
ringing this still leafless maple tree,

9

its branches only a wintry gray,
no sign yet of the sap of Spring,
the flow of new life. Three massive pines
evergreen impose color and shape in our line
of vision. It is enough for now, until
April rains descend and release
the green energy locked in the ground.
Then a familiar rejoicing will resume!

Dear friend, it is better, far better,
to have a heart broken than lost.
A broken heart still occupies
its niche in your breast, and casts
its broken light over body and soul.
That light, however imperfect,
is still divine radiance that ensures
the beauty of each passing day.
And beauty will attract other beauty
without end. It is only required
that you love the Earth, because
she is the Mother of us all,
she is hearth and home, source
and destination, the place
our children will inhabit...
Do not think to possess the Earth,
become one with the Earth. Make
of her abundance a daily feast
of never-ending proportions,
make of her blessings a place of perfection.

Mayjorey Buendia

Gaia

Hard cry, weeping thunder

Do you hear it?

Storm is coming, wind howling

Do you feel it?

Beginning of destruction, end of evolution

Gaia is mourning because of Global Warming

Stop the industrialization...

People and animals, we're all hurting!

Hand in hand we can solve the riddle

Stop the pollution, lets win this battle

Clean and green, will solve the puzzle

Listen to nature, lets fight the hassle

Fabrizio Frosini

Anthropocene Epoch

The Holocene has terminated.

In the timeline of Earth history

Human activity is now the dominant influence on

Climate & Environment. We are still within the

Quaternary Period of the Cenozoic Era, but the

'*Age of Humans*' is the new Epoch and our

Remnant plastics will be found for

Millions of years, everywhere, in marine sediments

Of this neglected planet. Until the plastics

Incorporated into the rocks, investigated by the

Intelligences who will discover the lost world, when

Visiting this arm of the Milky Way Galaxy, will help

Them to understand how much limited our flaunted

Wit actually was.

Fabrizio Frosini

('Anthropocene Epoch', Italian version)

Anthropocene

L'Olocene è terminato.
Nella cronologia della storia della Terra è ora
L'Attività Umana l'influenza dominante che plasma
Clima e Ambiente. Siamo ancora nel Periodo
Quaternario dell'Era Cenozoica, ma l'
"*Età degli Uomini*" è la nuova Epoca e i nostri
Residui di plastica verranno ritrovati per
Milioni d'anni, ovunque, nei sedimenti marini di questo
Bistrattato pianeta. Finché la plastica inglobata nella
Roccia, studiata dalle intelligenze che scopriranno il
Mondo perduto, allorché visiteranno questo braccio di
Galassia, li aiuterà a capire quanto limitato il nostro ostentato
Ingegno in realtà fosse.

Alem Hailu G/Kristos

A lake's obituary

The mob, elites, journalists
As well as poets like I
To our environmentally–unfriendly bent
Turning a blind eye
Also tardy in asking "Why
We strip of mother nature's green mantle,
While to maintain the statuesque
It gets locked in a severe battle?"
Equally not checking overgrazing,
We allowed fertile soil and sand
Amok, wild floods ride
To a close by touristic lake,
Whose mouth an expansion
Used to make
As much as its foreign body intake.

Soon, with the vast array of
Flora and fauna it supports,
Before we knew it

The magnificent lake died

Ceding place to a barren land,

An eyesore that looked a dump yard!

Author's note:
We used to believe the talk about environment change was a far-fetched prognosis but things began to change before our eyes. Seasons simply bear their names their features are completely changed!
This poem is dedicated to Haromaya Lake.

Nosheen Irfan

Earth is Crying

Earth, our planet is changing
Uncertain of her future
In a limbo, she is hanging

The onslaught is relentless
Industrialization, deforestation
Rapid as wildfire, happening all around
Concrete in place of trees
Artificiality, our creed

How long can the earth bear it

Stripped of verdure
Exposed to burning sun
She must be flinching
As Ozone is thinning

The air is invaded
The sea is violated

Fumes, gases—

Chemicals, toxins—

The monsters we have let loose upon her

Sucking her blood

Eroding her spirit

Precipitating an Apocalypse, are we?

Afrooz Jafarinoor

The Last Days of Planet Earth

(Elegy Written To Exist In Case Poets Become Extinct)

Winds are blowing over ruins

Clouds wandering round in a daze

Merely twisting together

Shedding a cold tear

On the soil dying out

While in agony to open

The last blossoms.

Dancing with the wind are the thorns

So that they survive a few days or more

By sucking on the soil,

Unaware that the soil's death

Is a preview of their own death!

How simply have they forgotten

The spring and the flowering plain!

Only the rocks long to reign

Over the lands going to ruin.

Only the wildflowers are weeping

Quietly mourning for the days coming.

Farzad Jahanbani

This Woman Is

This Woman is

Plastic

Hot

Nude

We are all faulty in her fall

With

Plastic wastes

Fossil fuels

Environmental disasters

And...

"Earth" is her name;

We all only watch that

Nude

Hot

Plastic

This Woman is

Srijana KC Rayamajhi

Wake up!

drifting iceberg, retreating glacier

there is still enough to chill our beer

mosquitoes heading north can't pierce our thick skin

garbage in oceans, on mountains, exceeds any bin

ugly seals scalding, polar bears going extinct

why should we bother

computing our carbon footprint?

we devour exotic foods travelling for miles

let's burn more fuel as we drive (not walk) awhile

sunny coasts, runny rivers, hither and yon we roam

all the while thermostats heat our home sweet home

mad weather, bad climate, the change we do face

why should we share?

better create war and famine so they exhale less

if ever oceans rise and apocalypse unfolds

we'll simply rule an underwater world

like Disney's King Triton, we'll grow fins and gills

we need to wake up before this fallacy/folly kills

plan, reuse, recycle, save and fear

drifting iceberg, retreating glacier.

Joji Varghese Kuncheria

Save the Earth

Who owns the Earth and the fullness in it?
Who can hold it in the palm of his hand?
No man or beast; no, none, but One,
The maker of it all, its keeper.

No eye has seen, no ear has heard,
Such marvelous works we see,
Works no one can fathom,
Yet we enjoy His majestic creation.

In excess, Earth's crust was drilled,
Soil and water resources wasted
Leaving cavities, causing earthquakes,
Bringing total havoc to life on Earth.

Fight a good fight to save the planet,
Make your sojourn as brave as ever,
Sacrifice to save our dwelling place,
To save planet Earth and mankind.

Let us love our planet to the core,
Which is given to us and our kindred,
Not to trample it underfoot as chaff,
Nor to make it all a concrete junk.

Trees, animals and the whole wildlife,
Even humans, destroyed and killed,
Many an innocent blood is shed,
Earth is crying out for peace.

Save the earth and bring peace,
Make it a harmonious place to live,
Save our planet and save the lives,
Make it a serene and tranquil place.

Tapera Makadho

Epitaph For Mother Earth

Earth's departure is at hand
through the wounds she bore
wave upon wave of conflicts
cowards of the new frontiers
to their wombs shall return

for whoever shall sow the wind
shall always reap the whirlwind
home is not without its hearth
nor the green to mother earth

now posterity must sit and listen
and never again cheer them on
they who claim to have power
see how they have killed mother earth
now a land of war and plunder

they are flexing their muscles
they are inventing gadgets

gadgets that emit caustic gases

and the oceans are filling up

but we shall always hold dear

the good memories we had mother.

Mallika Menon

Woes Of Mother Earth

When toleration reached saturation,
sorrows made her mood sombre.
Mother earth's soliloquies gushed out
like rushing water from a crack in a dam.
"I, the wonderful home planet
took birth a billion years ago.
Floras, faunas, human beings
mountains, mighty sea and rivers,
though all of them are my own pals
how come humans became foes of
mine to torture me in umpteen ways?
They burnt and cut many trees
for their own profits and benefits.
I, the green house turned barren and
vanished pastures, forests forever.
Then rain said goodbye, rivulets went dry
Birds lost their shelters and
many of them got endangered.
No cuckoo or nightingale would come

to wake me up or put to sleep

if the same status quo continued.

Atomic warfare and bombing

made me sick and I lost my charm.

Their innovations and inventions

though superb, caused damages like

global warming and climate change.

They littered me with plastic debris and

that made me breathless and helpless.

None took notice and I was sure that

same torture would last till my death"

Barry Middleton

Eden's Rose

once a garden
now an enemy
her tears are exhausted

the heat
the storms
the wars

drought and starvation
angry seas rising
terror and blood

no food to eat
no water to drink
no love

the beauty of her rose
is forgotten
lost in the universe.

Leloudia Migdali

Time to protect this Earth

Time to ponder on the edge of the precipice we stand
What kind of life on this world should we plan?
Do we endure suffocation from all kinds of harm
Or simply enjoy fresh air and emerald land?

Time to erase this dark endless list of Earth abuse
Gas, plastics and chemicals people produce
Contaminate, pollute, destroy rainforests
While debate over optimal solutions are sorest

Time to reverse the destructive human deeds
Soil damaging acid rain, invasive weeds
Choking skies that clog Earth's lungs
Fouled rivers, tainted birds, oily dumps

Time for mankind to understand
Earth's precious sky, sea and land
Show respect for its magic, be a friendly ally
Otherwise, all life will soon vanish and die.

Bharati Nayak

The Earth

Oh, Earth
You are shedding
tears in silence
As you see
Your children
So apathetic
To the pains
Of your anguished heart.

You are stunned
By men's bizarre actions
Uncaring to your woes.

As you suffocate from
toxic effluents
Your skin burns from
poisonous chemicals
and nuclear radiation.

From your lap
go vanishing, the sweet streams
and heat of the scorching sun
kill your beauteous green.

Birds and animals die
As they lose their habitat
Sea rises, rivers flood,
Ice melts from snowy caps.

The foolish man
When will he realize
That his reckless actions
Ring the death knell
For this beautiful planet?
Earth, our Mother Dear.

Madhumita Bhattacharjee Nayyar

For the Love of Mother Earth

The earth by us, badly raped and mutilated,
Badly injured and serrated,
What if her pain and wrath she demonstrated
To us, in her own way either extirpated
Or in her tears we are completely inundated,
And in turn decimated and deprecated!

Badly burdened the Earth shakes,
Many a homes in the process she breaks,
Many lives are bruised and battered,
And many a lives are then shattered,
But it is not her who is to be blamed,
It's us who are to be shamed.

Are we ready to bear the worse,
Prepared to bear her curse,
Can we lend an ear to her complaints,
Can we stop the constraints,
Practice some restraints,

Treat her with immense respect,

Make with her some human connect?

Nature has to be shown some love,

The Earth needs to be pampered like lady love,

To take care of all her elements,

We know she is not malevolent,

Still not show our arrogance,

If we need to partake of her benevolence.

We have to rein in our lust,

Lest all means we exhaust,

To build her trust in us,

Stop all our acts so ungenerous,

If we want to leave a beautiful Earth for the future generations,

Let's work towards creating colorful, airy, fertile nations,

Leave behind an excellent, aware civilization

That exists with Mother Earth in synchronization...

Valsa George Nedumthallil

Planet Earth

Though a tiny speck in the vast cosmic framework
Planet Earth is the most beautiful of God's handiwork
The one and only known habitation to all creation
Graciously bequeathed to generation after generation

Poets sang paeans, celebrating her charm
All creatures thrived well knowing no harm
The songsters sang melodies from every tree
The beasts freely roamed the forests in glee

Our Earth that once shone in all her pristine glory
Has now become a dreary, desert territory
All her wealth is looted by man's endless greed
To stop this ravage, she does silently plead

We prey on her as on a corpse by a ravenous vulture
Hoarding of wealth by all means has become our culture
We rape, we pillage and ooze out her blood
Suck her life sap to the point of casting her dead

Earth is now sadly turned into a vast garbage can
With rising pollution dimming her original glean
To man and beast it slowly turns into a living hell
In all attentive ears falls her resounding death knell

All creatures are moving slowly to their doom
With rising carbon, soot and yellow fume

Many species face imminent extinction
And life moves fast to a critical junction

Know, only through Earth's hearty grace
Survival and sustenance are possible for human race
'Save Mother Earth'! 'Don't leave her a grave yard'!
This is the soulful prayer from a grieving bard.

Mohammed Asim Nehal

Your Generosity and Our desires

O' Mother, the bearer of all burdens

We are your innocent children

In our lust, desires and dreams

We knowingly or unknowingly

Trouble you time and again

No creatures dwelling on you

Are more unjust than us

We use, we utilize and explore

Yet our unlimited wants never cease

Your warming is a signal

Yet we ignore it

Driven by desires

And ruled by wishes

We see but do not understand

You keep reminding

By tremors, quakes, floods

And we cry for our losses

Indeed, we have become selfish

Lost the sense of eco-balance

Deforestation, concrete jungles

Roads, bridges we make on you

Fuels we extract and emit on you

Yet you are generous with your produce

Let the wind tell us your agony

Let the clouds make us understand

Let the flowers remind us of pain

Let the birds sing your praise worthy songs.

Anil Kumar Panda

The Simmering Earth

Lonely is the day lonely is the night
The earth seems deserted, not a soul is in sight

Flowers stop to bloom, springs cease to flow
Trees are leafless, high peaks cease to glow

Pastures are barren, cattle die in the shed
Sheep famish in the fields, dogs run mad

Sparrows fall flat, crows stop flapping wings
Like a group of mourners parrots sit on strings

Days have become hot, nights vomit fire
Streets are silent, flies whine inside the byre

Feet get burnt, heads are going to burst
Fronds droop fast, ponds cease to frost

Silence rules thick, foxes crowd the streets

Corpses litter around rotting in the heats

Clouds cease to float, shunning the sky
Mirages abound fields, canals turn dry

Men behave wild, ready to eat the earth
Eyes are hollow sockets, empty is the hearth

Heart is heavy with loss everywhere there is a fear
Hope of survival is gone as if doomsday is near

Marcondes Pereira

Unbearable future

A coliseum of death will come.
Life will be harder than before.
No, we cannot ignore Earth's cry anymore.
We will destroy the living creatures home.

There will be less maize, water and rice.
Big floods will be stronger and stronger,
Heat-waves will bring brutal danger.
We'll gamble Nature's soul with a mad dice

Seas of trees will turn into dry pages
Bringing tears to all living creatures.
We will erase Planet Earth's treasures.
A crowd of fools, but we think we're sages

Sarah Persson

Earth's Survival

The vastness of this planet that is Earth,
In abundance it supplies with gifts for life,
It fulfils our every need and never questions,
And yet selfishly we disregard its plight.
We drive our fancy cars emitting fumes,
And bad chemicals we spread amongst our crops,
Our factories fill our skies with toxic gases,
Can we not get in our heads this needs to stop.

And the fighting that goes on between our nations,
Killing carelessly our mothers who give life,
Do these humans not support our worldly welfare?
Surely this should be a cause to end this strife.
The poaching of our wildlife still continues,
To provide the greedy man with precious wares,
But soon these creatures numbers will be zero,
Help us teach these filthy hands that do not care.

The ozone in our sky helps maintain balance,

Each and every living creature plays a part,
Our Oceans, seas and land we need to care for,
Earths survival rests on us to play it smart.

Sajee Rayaroth

Timeless In Time

From my garden to your graveyard
There exist a myriad stars of a summer night,
Beneath the moonlight I hold you
Before your blue eyes close in trance.

From mother's womb to the coffin
Million ripples travel through us,
The vibrant colours of youth
Fade away when we stop to dream.

Bound within a flowing lifetime
We sailed in the blue ocean as lovers,
Obsessed in your beauty and elegance
I feel the eternal rain and snow.

Days are like a moving tunnel
Today when I walk past this barren land – once my home,
Fighting for my breath in the swirling wind
Your whispers are lost unheard by the mankind.

For those who travelled before us
You were a Blue Planet of Life,
For those who will follow us soon
Who can save you to be radiant in their lives?

From my garden to your graveyard
Before those stars fall down from the sky,
Before your tears evaporate in the night
Accept this bouquet of love, timeless in time.

Marianne Larsen Reninger

Work for the Night is Coming
(old hymn)

Workers of the world unite
Before the nectar-laden world takes flight
And shreds our delicate web of life
Into Oblivion...

Forty Five days is all you've got
To feed the flock and pollinate the lot
Of Sweetest bee balm, lavender, and forget-me-not
For sacred honey...

Stingerless, dreaded Drones
Just lie around and groan
In ecstasy waiting for their turn
To pleasure the Queen...

She, the omnipotent one, spews her eggs
Commands the hive, creates the drive
To ensure the lives

Of her colony...

They tell me little worker bee, that if you cease
And night descends, and you and your sister's
Production ends, we Humans have but
Four more years...

Our world turns lonely in this vast universe
Homo-sapiens but a dot in Time's bio-diverse
Of creatures large and small.
But cease your life's work Ms. worker bee
And the world might retreat to the edge of the sea,
Revolve to the beat of stingray's tails and jellyfish swales...

Kirti Sharma

The Tree

Over the years, I've witnessed it all,
the drying up of water beneath my feet,
the level of rising heat above my leaves,
and the dying humanity of this century.

With every leaf that falls down,
I feel the warmth, not required now.
The rain, the breeze, those last few years I miss.
The pain of scars on my trunk,
well, this life was never my wish.

Their efforts to reach to the sky,
the 'development' has taken our homes.
Even we wish to see the stars at night,
if only their greed was less, to know.

The dust has choked our throats.
Threatenings are served to our souls.
Where in universe should we shift to?

They are eating away all the new resources!

The existence of life, is in question.

No way to reverse the destruction.

Everyday a life form dies.

Today a tree falls,

tomorrow these humans will cry.

Anzelyne Shideshe

Father Nature

Nature's magical existence

Everything at its core setting

Flourishing to decomposing

For billions of years

Mother Nature has been resilient

Surviving meteor bombardment

Worldwide volcanic eruptions

Even planet collisions

Plus several mass extinctions

Lots of life eliminated on the planet

Earth has depleted its pride

By pollution, biodiversity and deforestation

Its lungs filled with rusty poisonous air

Its heart hazardous by fluoro bi carbides

We measly human beings worried the repercussions

Perhaps the mother is in complete control

Ready to shade her skin

Eventually will all be gone

Devoid of father nature

Pamela Sinicrope

To the Polar Bear

I'm sorry poor bear-
160 days on shore—no ice,
no seals-no meals: that's hard.

When your skin hangs slack-
no fat, tired feet—and you find
a grizzly through tinder

wondering way too far North—
we'll get a Pizzly. I guess Nature
is evolving you off the species list.

And while I feel sorry for the seals
you consume (also very cute)—
I feel bad that consumption

of fossil fuels leaves you hungry. I never thought
my country would elect a sideswiping
blowhard who rejects global warming—

rejects you-King of the polar ice caps-largest,

most regal of bears. As ice parts,

more species become endangered-extinct.

One day our children will look out—see

a hollowed landscape—empty.

Too many (you) will be dead.

Future generations will power

computers-phones-brain chips—discover

Earth was once a bountiful garden.

Forgive me for what I did not do.

Kim Solem

I Believe

As defined by her defilers

The Earth must be a woman

These prolific greedy insatiable pigs

They suckle and bite her teats

Till they bleed, dry and wither

Corrupting her fertility

Pillaging her orchard green

Of every pollen and seed

Strangling her ovaries

Till she becomes a Cemetery

They taint her fragile womb

With chemical infection

Turning her sapphire seas

Into stagnating estuaries

A rotting pool of doom

Pious priests of hypocrisy declare

God is a man, not a woman

There is no Mother Earth Deity

Just terra firma for human greed

A spoil of man's war with nature

I say,

If Earth is not a Goddess

She's God's beloved mistress

And he'll punish her defilers

With such profound vengeance

They will perish from the universe

Forever and ever,

Amen

Udaya R. Tennakoon

Wonderful creatures!

I'm a bird, flying in the sky
Free and gentle as a cloud
When my wings are heavy and feel me hungry
I perch on a tree to look for my meal

Wonderful creatures!
The song I sing they don´t hear

They are living underneath me,
Not yet I realize what they do
One thing to be sure, I would feel
They are living in fear and in danger

Wonderful creatures!
The song I sing they don´t hear

Gestures, moods and vehicles
Sounds, fire, and massive things in horror
Same in shape but different in action

Everywhere the creatures in fear

Wonderful creatures!
The song I sing they don´t hear

Than they achieve, destroy the earth
Land they occupy never ever satisfy
Plan for life they have no clear
Sand and dust their hope might be near

Wonderful creatures!
The song I sing they don´t hear

Tze-Min Ition Tsai

The sunset even feel cold

That tide

infested waywardly my sandy beach

Sunset's advice

With red eyes

No day to let off

In the past ten million years

Those ungrateful westerlies

Always secretly come and also secretly go

To turn

The giant fan of that wind power tower

For the confrontation between man and nature

Do not say a word

Packed up

the nets that were hanging around

Do not understand in heart

How to deal with the questions of the little fishes

that desolate wind

Those thin meshes can arrest that cold

before jump into the sea

Savita Tyagi

The Dragon's Mouth

Our earth is a celestial gift of amazing notion.
Here exotic birds chirp in green meadows,
And frothy waves dance in blue oceans.

Fog and floating clouds shroud
Its snow-covered mountain house.
Their mystical silence so arresting
Chains the mind into a quiet standing.

Here life sings sweet and sour hymns.
In its lap we breath, sleep and conjugate.
What a sinister offspring we have become.
In its heart we have thrust a dagger deep.

A monster child we are, wrapped in greed,
Sucking lifeblood from our mother's breast.
Its lush green forests are becoming a graveyard
Of dead trees and an ecological disaster for rest.

Its waters we have polluted with chemicals.
Smoke filled chimneys spew our black hatred.
Plastic, carcinogens and poisonous gases
Fill our land with filth to tarnish its acreage.

Its grace is crushed under sky rise buildings.
From the smog filled sulfurous sky,
The Sun's ultraviolet rays pour through
Broken shields of ozone, send us the chilling vibes.

Our Mother Earth is sending us clear signals.
We need to understand its angry outbursts.
Those falling glaciers are flooding the land and oceans.
The chemical discharge is scaring pristine rivers' bosom.

Oil and coal polluted ether is changing the earth
In a superheated dragon's mouth of leaping flames.
Pushed to the edge it would burn and devour us alive.
Our beautiful planet is in grave danger and we are to blame.

Hans Van Rostenberghe

Fever!

To our dear mother earth,
who gave to all of us birth:
your fever we want to cure.
Will we succeed? Not so sure!

Some of our sisters and brothers
seem not to care for any others.
They seem not worried about your health,
apparently blinded by money and wealth.

When pushed on the topic of your fever,
they will make any kind of endeavor
to deny it and claim nothing is wrong;
a lack of emotion, their greed much too strong.

I still hope, our dear mother earth,
holding all treasures of incredible worth,
that we will find people in numbers enough
to treat your disease with utmost caring love.

Let us work all together to find a solution,
stop all the waste and unnecessary pollution,
reduce carbon emission, preserve our trees,
keep clean our earth, our oceans and seas.

Now is the time to move and act.
The disease of our mother is a certain fact
No man with a bit of intellect, a bit of a brain
can be allowed to ignore your fever again.

Joey Vlahakis

Species

Are humans selfish
Or at least more so than other animals
Would another animal do the same
If they were the dominant species

As a dominant species
Do we have the obligation
To care for other species
As a king

Or do we continue
As dictators
Who put our needs
Above other species

If we cannot even unite
Behind the many different colors
How can we act as a king
Who is fair

Have we missed our purpose

Instead of being successful

Should we be caring

And kind

Do we apologize to Earth

And her children

For our behavior

Or do we remain the spoiled child?

Let us build a ship together
and call it World Mercy
and steer it towards
the sunrise.

Diana van den Berg, 'Africa Mercy'

Contributors' Biographies

Alexandro Acevedo Johns, Chile

— My name is Alexandro Acevedo Johns, but I sign my writing with my maternal surname (Johns). I am Chilean, born on November 2, 1947. I'm a lawyer and live in Santiago, the capital of Chile, with my wife Marcela. In my youth I was devoted to poetry, as many of my generation. Now, since I retired from the legal profession, I've regained my freedom to write. It is said that writing is a very demanding activity and endanger the spirit if you're not an optimist. But, after the years, I feel that writing helps me to stay alive and connected emotionally with the world we live in.

~*~

Ellias Aghili Dehnavi, Iran

— I was born in 1996 in Iran, and I'm currently living in Esfahan, the cultural capital of Iran. I'm studying English literature at the University of Esfahan (B.A student). My favorite fields of study are poetry and English literature. I wrote my first poem, a limerick, when I was twelve years old, and compiled my first Poetry collection, on peace as a topic, when I was 15. One year later, this poetry collection got a recognition from the faculty of foreign languages (University of Isfahan/Esfahan), and also hit an important festival in Iran, called "Khawrazmi". Since then, I've published some other poetry books, also with friends, members of the M.O.P international group, of which I'm currently the second secretary. Since we are all seeking for a better world, where peace and friendships are basic values, it's a honor to be part of 'Poets Unite Worldwide'.

~*~

Saadat Tahir Ali, Pakistan (currently in Saudi Arabia)

— I was born (in Jan. 1965) and bred in Pakistan. A medical doctor by profession, with postgraduate qualifications in Radiology, I'm currently living in Qaseem, Saudi Arabia. My hobbies include indoor plants, interiors and woodwork.. and making friends. Over the years, I have traveled to many countries and as a reasonably experienced traveler, I am a senior reviewer on travel and foodie sites. I like nature landscapes architecture and history. I am averse to concrete jungles. I am a diehard audiophile. I consider myself a wide eyed student, ready to listen, learn and improve. I loved poetry when I was at school, started writing decades back while at cadet high continued through to King Edward Medical College. Freedom from bondage in all forms and colours, love and universal brotherhood are my cherished values. I am an incorrigible romanticist and love music. I write my mind..

~*~

Anna Banasiak, Poland

— Born in Poland in 1984, I live in Łódź, in the central part of the Country. I'm a poet and literary critic. The winner of poetry competitions in London, Berlin and Bratislava, my poetry can be found in most of the Anthologies published by F. Frosini & Poets Unite Worldwide. I'm interested in Art and psychology.

~*~

Tom Billsborough, UK

— I was born in 1943 in Preston, England, and currently live in Kirkham, North Lancashire. I'm a retired chartered Accountant. I write poetry in English & French, and translate from French & Spanish.

~*~

Daniel J. Brick, USA

— I was born in Minnesota, in the Twin Cities, in 1947 and lived my whole life here. This is where I am rooted, near the Mississippi River, in a landscape of four seasons with many trees and parks and lakes. These are the natural things I treasure. Poetry and classical music are my passions. Over the years most of my friends have moved to warmer climates, so in old age I find myself to be something of a loner. But I have a talent for solitude. A good number of my poems are published in the book "The Double Door", written with Fabrizio Frosini.

~*~

Mayjorey Buendia, The Philippines

— I'm Mayjorey Dellosa Buendia, from Mandaluyong City, Philippines. Born in 1981, I am a graduate of Business Administration, Major in

Management. I find solace and peace with nature, arts, poetry, reading and writing. I'm also into Women's Empowerment & Independence.

~*~

Fabrizio Frosini, Italy

— Born in Tuscany in 1953. Currently living close to Florence and to Vinci, Leonardo's hometown. Doctor in Medicine, specialized in Neurosurgery, with an ancient passion for Poetry. Author of over 2,000 poems, in 16 collections. Thirteen of them are also published as ebooks —among them: «The Chinese Gardens», «Prelude to the Night», and «Karumi – Haiku & Tanka» Author's Page at Amazon: *https://www.amazon.com/Fabrizio-Frosini/e/B014HA8ZUA/*

~*~

Alem Hailu G/Kristos, Ethiopia

— Since in Ethiopia we use our age-old style to name a new born child (chosen Name, Father's Name, Grandfather's Name): Alem is my name, Hailu is my father's name and G/Kristos is my grandfather's name. Born in Ethiopia in 1974, I live in Addis Ababa, Ethiopia capital city, where I'm currently deputy Editor-in-Chief of the Ethiopian Herald. M.A holder in literature, from Addis Ababa University, I'm a published poet, novelist, editor, translator of masterpieces, literary critic, playwright and journalist. My book "Pupils' Poems" has been published by Lulu (ISBN 978-1-329-30770-4), and my novel "Hope from the Debris of Hopelessness" by Noah Books (2017, ISBN-10: 194811707X).

~*~

Nosheen Irfan, Pakistan

— Born on 13 March, 1978 in Lahore, where I currently live. I studied English Literature at the University of Punjab. I teach English to secondary classes. I'm the daughter of a civil engineer who inculcated the love of books in me. Thanks to his taste in literature, I had access to some great literary classics at an early age. I became an avid reader and have gathered an impressive collection of books over the years. I turned to writing at a later age but have made it a point to write daily ever since. I draw inspiration from both classical and contemporary literature. Apart from that, Nature, people, life and social issues inspire me to take up the pen. I have my work published in "Eastlit" magazine and hope to find a wider readership in the future. I have had the good fortune to read great literary works by internationally renowned writers that have enlightened my mind and broadened my horizons. Literature has helped me grow as a person. I have become more open-minded through my exposure to great minds and have gained a broader perspective on life. I will love to take up writing as a profession if the opportunity arises. Yet, at the moment I'm content writing to express myself creatively. It gives me the greatest pleasure to become a voice that is heard somewhere.

~*~

Afrooz Jafarinoor, Iran

— I'm a teacher and writer, living in Tehran. I was born in 1972 in Hamedan, Iran, from lower middle class parents who loved books and raised me with lots of fairy tales. Learning Persian, Turkish, and Kurdish as a child has given me a passion for foreign languages. I hold a master's degree in English literature, and have also learned a lot of German. I write poetry in Persian and English and also translate into them. Another field I love is theater and cinema. I also hold a master's degree in dramatic literature and I write reviews. To publish my poetical compositions and translations, as well as my critical essays, I launched a website: *www.poetsrepublic.com.*

~*~

Farzad Jahanbani, Iran

— I was born in 1980, in I.R. Iran, and live in Tehran. I've a master degree of EMBA. I have been writing poems since 1995 in my native language (Persian), then also in English. My poems are in several Anthologies edited by Fabrizio Frosini and Poets Unite Worldwide. See me on Telegram messenger *www.telegram.me/Farzadia.*

~*~

Srijana KC Rayamajhi, Nepal

— I was born in 1990, in a middle class Hindu family, in Kathmandu, where I live. I have two siblings. Being my father a Captain in Nepal army, I went to the army school. I am a doctor by profession. I write both to express myself and for my love for Art and Creativity. Besides I like painting, reading books, visiting places, trekking at times. Yet, sleeping is what I'd like the most, if only I could.. I try to listen to Buddha's words and therefore pursue peace since childhood. My name, Srijana, in Nepali means 'Creation'. My poems are "born of heart", *Manasija*. My poems are "born of mind", *Manoj.*

~*~

Joji Varghese Kuncheria, India (currently in Oman)

— I'm an Indian national working in Muscat, Oman, since 2004. I was born on February 14, 1953, in India. I did my M.A. in English literature from Christ Church College, Kanpur (Kanpur University, India). I'm a senior Lecturer, teaching British and American literature to the undergraduate students in a college in Oman. I started writing poems while I was working as a teacher in Ethiopia (1978–1985) and have continued to write, after a long gap, from 2009 onwards. I'm very passionate about the peaceful co-existence of the people anywhere in this planet, and cherish to see such a world order. I'm a good chess player too.

~*~

Tapera Makadho, Zimbabwe

— I was born on the 18th of February 1974 in the communal lands of Zaka District on the periphery of Masvingo in Zimbabwe. I attended Mafuratidze Primary School from 1982-88 before enrolling at Machingambi Secondary School where I attained my GCE 'O' Level in 1992. I lost my father in 1988 and grew up under the care of my grandmother as such advancing with my education became difficult. Pursuing higher learning was a luxury they couldn't afford hence I had to contend with my GCE certificate. In 1997 I trained as a Policeman with the Zimbabwe Republic Police whereupon completion of training I was posted to the quasi military wing of the Organization; Support Unit. Subsequent to my posting to this Unit, I found the duties there challenging, at times dictate that I preoccupy myself with writing especially when deployed in the scary game parks. Spanning to over 18 years, these grueling experiences undoubtedly carved both my character and the manner in which I perceive life issues. I then started writing poetry sometime in 2005 after a former colleague and close friend, Elvis Nikisi who was already into poetry had invited me to write a poem which I did and have been writing ever since. I'm married to Caroline and we are blessed with a daughter, Kudzai.

~*~

Mallika Menon, India

— Born on the 23rd of January 1961, in India, I hail from Kerala's capital city, Trivandrum, on India's southern tip, but I enjoyed my life in Mumbai. Lover of music and literature, I sing songs and poems. One day, I started singing my own poems! I offer collection of poems in mother tongue Malayalam as well as English. Simple emotions, gentle feelings and shades of empathy reflect in my poetry. I like reading philosophy. I enjoy interior decoration. I'm travel-savvy, keen to explore cultures and cuisines world-over.

~*~

Barry Middleton, USA

— I was born in 1946 in a small town at the edge of the Mississippi delta in the United States of America. The town and surrounding countryside would have been great subject matter for Norman Rockwell paintings. Much of that landscape and many of my experiences there are reflected in my poetry. I was educated in the parochial school system in Yazoo City, Mississippi, and graduated High School in 1964. I attended Spring Hill College in Mobile, Alabama, and graduated in 1969 from Southern Illinois University in Carbondale, Illinois, with a BS Degree in Psychology. After graduation, I worked as a school teacher in the then segregated African American school system in my home town and later as a social worker in Orlando Florida where I worked with all kinds of troubled and abused children. In 1987 I graduated from Rollins College in Orlando with an MA Degree in Counseling. For the remainder of my career I worked as a Licensed Mental Health Counselor at Venice Hospital, in private practice, and as a group specialist at Sarasota Memorial Hospital. I currently reside in Sarasota, Florida. The most important accomplishments of my life have been watching my son grow up and having the opportunity to fulfill a lifelong dream to write and especially to write poetry.

~*~

Leloudia Migdali, Greece

— I was born in 1959 in Itea, a nice little city close to Delphi, 'the center of world'. Attended school there till 1979, then followed a course in the English Literature Department of Aristotle University of Thessaloniki,

Greece. Back to Itea, where I still live, ran my own English institute till I was appointed at the public sector. I have been teaching English for the past 29 years, in Primary, Secondary and High school as well as in the Maritime College in Galaxidi city. Meanwhile I got a postgraduate degree on Teaching English as a Foreign Language from Patras University. Poetry and writing has always been my favorite hobbies. After my retirement from public sector, I have been devoting more time in writing poetry. Currently writing poems on life and the way I see it. I also write contemporary articles on online sites. Happily married and mother of two children.

~*~

Bharati Nayak, India

— I hail from Odisha, an eastern state of India which has a great heritage of art and architecture. I was born on the 26th May 1962 and live in Bhubaneswar, capital city of Odisha. I have a Masters Degree in Political Science from Utkal University, Vani Vihar, Bhubaneswar, and have a Government job. As a bilingual writer (Odia and English), my published works include a devotional poetry anthology in Odia, 'Padma Pada', and a collection of poetry in English, 'Words Are Such Perfect Traitors'. My poems have been published in many newspapers, magazines, journals, e-books and poetry anthologies of national and international repute, such as 'Rock Pebbles', Odisha Review, Utkal Prasanga, Creation and Criticism, Literary Herald, Glimpses, 'Splash of Verse' and the like. Poems of mine are also in many Anthologies published by Fabrizio Frosini & Poets Unite Worldwide. I have a blog, 'Bharatispen', at Wordpress. I also take interest in social issues.

~*~

Madhumita Bhattacharjee Nayyar, India

— Born on 15th of April 1978, I live in the NCR of Gurgaon, in the Indian state of Haryana. I have been writing since childhood, churning my imagination wheels. After completing my masters in English, I edited and wrote for MEA, MDU and others. I started my career with media (television and radio), moving onto the perfumes and cosmetics sector, writing product stories, designing etc, gradually moving onto healing and life skills counseling. I am very sensitive to whatever happens around me and give vent to all my feelings through words, both in prose and poetry. I write both in English and Hindi. My works have been published on some portals and anthologies. An avid animal lover, I believe we all can do our bit to bring in happiness to the world. My motto is "Live and Love Life".

~*~

Valsa George Nedumthallil, India

— Born in 1953, I live in a suburb of Ernakulam, Kerala (in the south-west of India), where I lead a happy and contented life. After a successful career as a college teacher, when I retired from service I took to poetry. Now it has become an obsession and a mentally rewarding engagement. I write on a wide spectrum of topics spanning Nature, Love and Human relations. I have to my credit four published volumes of poems: 'Beats', 'Drop of a Feather', 'Entwining Shadows' and 'Rainbow Hues', the latter two available on Amazon.com. As most others, I long for a peaceful world where man is bound to man by the invisible thread of love and live in amity and harmonious co existence. In a strife ridden world it is incumbent on the part of a poet to dwell upon social issues like war, corruption, poverty, refugee influx etc.. We should hold a mirror to life and its sufferings and joys and strive to bring each one's life a trifle closer to the worth and meaning of man's existence on this Earth. Only by exposing the injustices and absurdities of society and voicing against them, we can reduce the gloom that permeates the lives of many in our society.

~*~

Mohammed Asim Nehal, India

— Born in 1970 and brought up in Nagpur, the orange city of India, I started writing poems and stories at very young age, as I was inclined to write something to express what I feel and it comes from the deep thickets of my heart. I feel poetry is a trick of language that magnetizes the readers and takes them to a world that is virtually created by the poets. A number of my poems have been published on some magazines and newspapers. By profession, I am a Chartered Accountant and a Company Secretary with Master's degree in Commerce and Bachelor's degree in Law. What I studied for and what I feel are two parallel lines.

~*~

Margaret O'Driscoll, Ireland

— I'm Irish, born in Sept. 1960; a mother of seven and grandmother of eleven. My poems have been published in various anthologies, ebooks and magazines. I've recently published my first collection of poetry, 'The Best Things In Life Are Free'. I'm a Social Care Worker and like walking, singing, dancing, and reading, to unwind. I love to spend time with my grandchildren exploring, gardening, watching films and cooking.

~*~

Anil Kumar Panda, India

— I was born on the 18th of April 1962, in a small town, Brajrajnagar, in the state of Odisha (or Orissa, in the eastern coast of India), where I'm currently residing. Our family was basically poor, and that was why I could not continue my study after my graduation in Physics. I liked science and wanted to be a scientist. Since I was about 10 years old, I had developed a strong appetite for reading and writing. Presently, I am working in coal mines and write stories and romantic poems whenever I get time. I love nature and the abundance of beauty it holds in its lap. During my school days I used to visit our village with my grandfather to supervise cultivation work. There, I used to loiter around the village and enjoyed nature: the deep greenery surrounded by undulating hills fascinated me, and I enjoyed the changing of colors of the landscape with changing of the seasons. Using 'Tiku' as my pen name, most of my poems are on nature and simple lives of my village folks. I have published a book titled 'Fragrance of Love', and I'm currently working on my second.

~*~

Marcondes Pereira Da Silva De Mesquita, Brazil

— Born in 1991, I live in Barueri (State of São Paulo, Brazil). I'm a poet who is searching for my own truth, in this liquid world. I write to understand myself and the chaotic universe we're living in. I love to study languages and listening to Heavy Metal. My poetry speaks about war, love, religion, Philosophy, History and several other themes, although I write chronicles, tales and theatre plays too. My biggest influences in terms of poetry are: Camões, Petrarch, Boccaccio and Homer. I study their texts to create my own epopees, which I would like to see transformed in music. I am a technologist in Human Resources (Faculty Fernão Dias). I love to learn about different cultures around the world, and I'd like to show the world my poetry and prose writings.

~*~

Sarah Louise Persson, United Kingdom

— I was born on 14th October 1966 in Wellington, Shropshire, England. I have lived in both South Africa and Denmark in the past, but have come to settle in the beautiful West Yorkshire, England, currently living in Leeds. I work in a bank but my true passion is in poetry. I used to write so much as a child but it faded into the background of my life until recently. I suffered a depression and the writing of poetry helped me get my feelings down on paper and eventually helping me to deal with my emotions in a positive way. I now write almost daily and am a very happy individual. The best thing is that I get to share my poems with such a variety of people from all walks of life and the feedback inspires me even more. A few of my poems have been published in Anthologies edited by F. Frosini.

~*~

Sajee Rayaroth, India/Australia

— I am an Australian Citizen living in Brisbane, Queensland and a Chartered Engineer by profession. I was born in India and hold an Overseas Citizen of India status as well. I did my Post Graduation in 1995 and currently pursuing PhD at University of South Australia. As part of my engineering profession, I have lived in many parts of the world including the Middle East, Europe, South East Asia and now settled in Australia. This gave me a chance to learn various cultures, people and way of life in different countries. I have an immense passion towards literature. I have written several poems and short stories mainly in two languages: English and Malayalam (Indian language). Being an avid reader and enthusiastic writer in the social media, I have published some of my works in National and International media.

~*~

Marianne Larsen Reninger, USA

— I was born in Denmark in 1944 and emigrated to the United States with my parents in 1947. I began painting and writing at a very young age and by 16 was studying painting and taking commissions. My prime influence was a Russian born artist, Tatiana McKinney, world famous with work in the Vatican and in major museums. From Tatiana I learned to see "the atmosphere between the mountains" and the "meaning between the words". Today, I paint and write from my mountaintop home near Asheville, N.C.. I consider myself an "editorial artist" with my acrylic/collages often containing original poetry. My work is textural, touchable glimpses of the natural world and my reaction to life's political and social merry-go-round, words become as important as the brush strokes. The work is meant to be read, like a favorite book or poem, as well as, absorbed, like a visual feast. You can see my work at Pinterest.com, Marianne Reninger, Marianne's Art Board.

~*~

Kirti Sharma, India

— I was born on 7th October 1996. I live in Delhi, India. Currently I am pursuing my Under graduation in Science from Delhi University. I started writing poems when I was 15 years old, as that part of my life was a turning point. I realized the true happiness lies within one self. The world of poetry became a part of my life and I started reading & writing more of them. I learned how each line of a poem has its own music. The poems written by P.B. Shelly are my favorite; 'Goodnight' being one of them. I generally write poems on love and solitariness as these two conditions are common in every person's life. My poems consists of simple words & are easily understood. My other hobbies are singing, dancing and reading novels. Paulo Coelho is my favorite author.

~*~

Anzelyne Shideshe, Kenya (currently in Germany)

— born and bred in Eldoret, Kenya, in 1982, where attended high school up to year 2000. In 2002 I moved to Mombasa where I studied marketing at the Technical University. Working in sales and marketing has been a plus to more experiences. I'm currently in Germany (Baden-Württemberg) and my passion for writing intensified.

~*~

Pamela Smith Sinicrope, USA

— I was born in Frederick, Maryland in 1970. I spent most of my youth in Texas; and now, for the past 13 years, I have lived in Rochester, MN, with my husband and three children. I have a doctorate in Public Health, with an interest in reducing morbidity and mortality from cancer through focus on families, lifestyles, and genetics. As a young child and raised as a Unitarian, I have fond memories of hiking in the woods and reading poetry while sitting on rocks and communing with nature. When I was twelve, my grandmother and I used to write letters and share our poetry with each other. I enjoy: tennis, music of all kinds, reading (poetry and prose), and spending time with my family. I returned to writing poetry over the past year in my 40's. Now that my children and I are older, I have time again, to reflect, write, and read. And I love doing it most every day!

~*~

Kim Alan Solem, USA

— A retired steel worker born in 1956, I live in St. Paul, MN. A lifelong lover of poetry, I first started writing only four years ago, after being diagnosed with stage 4 cancer. I needed a distraction from the challenges of chemo and radiation treatment, and writing poetry gave it to me. In my third year of remission, I can attest that writing poetry can indeed be therapeutic. I plan to keep on living by keeping on scribbling. When I'm not writing, I spend time with my four wonderful nieces, whom I say "are driving me to pieces."

~*~

Udaya R. Tennakoon, Sri Lanka (living in Switzerland)

— My full name is Udaya Rathna Tennakoon Mudiyanselage. As a Diaspora Poet, I live in Zürich, Switzerland, but my home country is Sri Lanka, where I was born in 1970. Being a political refugee, I could see the world in many perspectives and engage with writing and research. I graduated from University of Colombo and University of Kelaniya, Sri Lanka. At the University of Basel, Switzerland, and also at the University of Innsbruck, Austria, I studied 'Peace and Conflict Transformation' for my master Studies. As a writer, I've written some theater works and contribute articles to many websites and also as a social activist, I engage with many voluntary works inside Switzerland and Europe as well as Nepal and Sri Lanka. I've published a haiku poetry book titled 'The Fragrance of Loss' (2017).

~*~

Tze Min Ition Tsai, Taiwan

— My name is Ition Tsai 蔡宜勳, where 蔡 (Tsai) is my Family name, while Tze-Min Tsai 蔡澤民 is my pen name. Born in 1957 in Taiwan

(Republic of China), I live in Changhua city. I hold a Ph.D. in Chemical Engineering, and a Master of Science in Applied Mathematics. I have equal affection in science, mathematics and literature; the results are all reflected in my academic and creative writings. I am an Associate Professor for Asia University, Taiwan; at the same time I am a columnist for several poetry journals as well as the editor of "Reading, Writing and Teaching" academic text for the National Changhua Normal University, Taiwan. My writing includes novels, prose, and poetry, and I specialize in describing nature and humanity's love and affection through these creative literary works (for such a reason I am often referred to as a "green poet"). In addition to receiving many domestic and foreign literary awards, a large number of my works have been translated into more than 13 languages in over 37 countries.

~*~

Savita Tyagi, India/USA

— Born in 1948, I was raised in India. As a student of liberal arts I loved history and literature and completed my M.A. in Western History. After marriage I migrated to California and later came to live in Edmond, Oklahoma, where I reside currently with my husband of 46 years. In U.S. for some time my love for reading was limited to English language. However the need to expose our children to their language, culture and religion brought me back to my roots. While organizing children's classes at my home and temple, I started studying more of our spiritual books, in English as well as in Sanskrit, and realized that some of the best poetry was hidden in the ancient literature. While ancient poetry takes me to the path of self discovery, the contemporary writings help strengthen the social consciousness. Some of my poems have been published in newspapers, anthologies and magazines. A self published book of poems 'Back Yard Poetry' is available through Blurb. But most of my poetry and other writings are on my blog 'When Thoughts Get Wings'. Besides reading and writing I enjoy walking, quilting, meditating, learning from Nature and visiting with friends. I am moderate in my views and have deep respect for human values that nurture and nourish us in all walks of life.

~*~

Hans Van Rostenberghe, Belgium/Malaysia

— Born on October 18th 1964, in Oudenaarde, Belgium, I'm currently living in a town called Bachok, in Kelantan State, Malaysia. I am a doctor in medicine (neonatologist) and a professor at Universiti Sains Malaysia, where

I have been working since 1994. Among the most important sources of inspiration in my life are Dr. Albert Schweitzer, Dr. Martin Luther King and the Organization 'Médécins sans Frontieres'. Poetry has become a passion since 2010, when I was bedridden for three months, due to a fractured vertebra: I write under the pseudonym 'Aufie Zophy'. I am a reader of philosophy, a nature lover and a family man. I believe strongly that the world is heading towards harmony through an ever increasing kindness revolution which is close to its sharp inflection point on its exponential curve. My blog: *http://reflectionsbyhans.blogspot.com/*

~*~

Joey Vlahakis, USA

— I was born in Sept 2000 in Rochester, MN and other than two years in San Francisco, CA, have spent most of my life in Rochester. I am now in the 10th grade at Mayo High School, where I am on the debate tennis teams. I follow soccer and am an avid Liverpool F.C. fan and enjoy video games and time with friends and family, especially my dog. I have always been fascinated by History and my favorite era is the Greek empire. I love to travel with my family and visit relatives abroad. I started writing poetry a few years ago because to me, it is the only fun form of writing. Poetry is the best way to tell a story and paint the picture of a scene.

~*~

... All night
I wander alone, searching in vain for the irretrievable.

Campbell McGrath, 'Nights on Planet Earth'

Poets Unite Worldwide

'Poets Unite Worldwide' represents, in my mind, an invitation and an appeal (*"Poets worldwide, unite!"*), and it is more an open group of poets, an independent community, than a formal association —but still an 'Association' of over two hundred free minds and spirits.

I'd say that this comes, first, from my own nature: I consider myself not just an Italian, but a Citizen of the World —born in Italy by chance—, equal to everybody else: all human beings on planet Earth, in brotherhood. I have an independent mind and the utmost respect for the human values of freedom, justice, privacy.. and I dislike almost all kind of formalities: for such reason I stay away from anything that sounds bureaucratic.

Although living in different countries and continents, we all feel a kinship, being part of this poetic drive for worldwide peace and brotherhood. In such a way, we work together for the highest purposes, as all mankind should do.

I can say that 'Poets Unite Worldwide' was born, in its extended form, in the Autumn of 2015, when I invited tens and tens of poets, worldwide, to join me in writing a poetry compilation on (against) terror, in response to the bloody Paris events of November 13, 2015.

I felt the urge, that time, to began working on a new ebook, 'Poetry Against Terror', and I enlisted 'my' community of poets worldwide to help, since I wanted it to become a large collective work: the voice of poets from many different countries, worldwide, who stand up and speak aloud, but without hatred, against the bloody madness of terror. Astonishingly, 64 Poets from 43 countries lent their pens in the effort, and I wrote, in the introductory note to the book, "we—poets of the world—wish to make our voices resonate in the minds and hearts of all women and men who refuse to be silenced by hate and violence." Pamela Sinicrope and Daniel Brick, both of Minnesota, USA, along with Richard Thézé, England, co-edited the collection of diverse poems about terrorism —in Paris and around the world. Cover art was by Galina Italyanskaya, Russia.

The project came together quickly, with poets coming from countries in all continents, including Arab/Islamic countries: *Australia, Bangladesh, Botswana, Brazil, Canada, Chile, China, Croatia, Egypt, France, Germany, Ghana, Greece, India, Indonesia, Iran, Ireland, Israel, Italy, Kenya, Morocco, New Zealand, Nepal, Nigeria, Oman, Pakistan, The Philippines, Qatar, Russia, Saudi Arabia, Serbia, Somalia, South Africa, Sri Lanka, Sweden, Switzerland, Thailand, Tunisia, United Arab Emirates, Uganda, United Kingdom, USA, Zimbabwe.*

Poem topics range from a focus on the liberty of France, to the musings of a mother who does not want her child suffering from terrorism, to a young woman who incessantly searches Google for the answers to the terrorism problem, to the story of African villagers who drink from a cow's horn under a peaceful moon until terrorism takes over. Many of the poets have experienced terrorism first-hand, and this witness is expressed in their writings and their biographies. As Pamela Sinicrope said, "We've all been touched by terrorism. For some, the topic hit home after the events in Paris, but for others, terrorism has been a disturbing part of everyday life — these facts are borne out in the poems. The poems speak for themselves."

Yet, as a group of poets collaborating together on a variety of projects, we didn't stop with that first book. We do have a website, that Udaya Tennakon created, as well as a Fb page. Since then, we've been continually publishing and growing, and –hopefully– improving as writers.

In Spring 2016 we published the ebook 'Poets Against Inequality', to add our voice to those other unequivocal voices that denounce an absolute lack of equality in our society. The poems collected in that book (as well as the previous one) belong in what is called "Poetry of Witness", and we believe that this is a task that all of us, as poets, have a moral obligation to pursue, because we can't accept to live in a world where extreme poverty is so widespread and sheer inequality is the norm.

Another project accomplished is a book on the Refugees theme: in March 2016, while looking at an image taken on the border between Greece and Macedonia, I felt the urge to write a poem. From that urge, a new editorial project was born, the book 'By Land & By Seas'; then others followed, like 'We All Are Persons – Why Gender Discrimination?', and 'Time to show up – Poetry for Democracy'. And surely, after the present book, thanking the enthusiasm and energy of many in our group, new good projects will follow. Our mission keeps on.

(Fabrizio Frosini, on behalf of 'Poets Unite Worldwide')

From a conversation with Ellias Aghili

— How did the idea of your first anthology sprout in your mind, Dr. Frosini?

It happened at the beginning of 2015: in mid January I proposed to my friend Daniel J. Brick, a poet from Minnesota, to publish a poetry e-book together. Daniel loved the idea, and in his answer he told me about a group of writers who published an e-book of English/Spanish songs.. that made me first thought about the possibility to add a number of poems by a few other poets from different corners of the world who'd like to join us.. and suddenly (while listening to the ouverture from Bach's Orchestral Suite No. 3) I had the "vision" of a number of Anthologies, on different topics, with poems from poets worldwide —by the way, this is why I used the term "Project Ouverture" to refer to that idea.

— Do you think that poetry can mark significant changes in the world we live in?

I'd love.. Unfortunately, it is not so. The contemporary world is too complex and poetry is not seen as an important 'voice' by the vast majority of the sheer minority of those who manage the so called 'political power' (in governments and other democratic institutions), but also by most of the so called 'public opinion'. One of the reasons, from my point of view —and I think that this is not a small issue—, is because poetry is not about money: it is not 'business'. In a human society where money is so valued, poetry is not considered of 'practical value'... It was not so in Classical Greece. It was not so in Classical Rome. It was not so at the time of the Tang Dynasty (China) and in Classical Japan... or just think of Goethe's thought of ancient Persian literature as "one of the four main entities of world literature", and how poets were highly regarded in those times. No more so today, worldwide —with just a few exception.

— What about 'Peace'?

'Peace' is a beautiful word, but a difficult concept for many human beings (well, I wish to think that they represent a "minority", but still too many!). The same can be told of the term "Democracy". A few months ago we published a book titled 'Time to show up – Poetry for Democracy'. The concept of "democracy" is quite fuzzy in many minds.. I have traveled the years and still don't know how people think.. What does "the many", "the People" —'Humanity'—, mean to them? Do people think to everybody or just to themselves? Generally speaking, they think themselves to be more legitimated "to live" and "to have" than the others. As a matter of fact, oftentimes people try to forget about humanity. Then a question arises: do people, ordinary people as we are, consider all to be equal in value? I'm asking, because this is the 'key' to Democracy —and is "key to peace" as well. Do people care of everybody else as they care of themselves? Or, to put it even simpler: do people think to everybody else or just to themselves? Many people are literally self-centered: they tend to think highly of themselves, and rate the others almost naught. There are people —and not quite a few— who do not think that each of us is part of the whole. They are simply "the whole" to their own eyes. The concept "if you damage one, you damage all" should be central in a true democracy. Unfortunately, it is not in the real world.

— *What do you think we, as humans, have omitted in our lives?*
The world of humans is complicated. The human society can be spoiled and even dissolved by its members: they could be a minority but, if strong enough, they can put our society at risk and, by weakening the democratic rules —day after day— can make society collapse. Human history should teach us all a lesson in humility. On the contrary, we try to deceive ourselves into thinking that what happened in the first half of the 20th century can't happen again. It is just a deception. If we don't understand that the human society is a living entity, and that we need to put 'love' into it, every day, so to keep it "vital" and in good health, the 'evil' inside us will destroy humankind.

— *What are your suggestions for the young people who wish to enter the wonderful land of poetry?*
Great.. come on! I regard Poetry as the highest form of Literature: you need a brain to write prose, but for poetry, you always need to add a heart, and they (heart and mind) have to talk to each other, being in tune and moving "hand in hand". We can't 'teach' other human beings to 'live', but we can show them how we live and how we write poetry.. if they are

90

interested in learning something about us and the way we 'make poetry'. The more young people embrace Poetry, the more it will let our hope in Humanity grow. Should we refer, as an example, to the topic "terrorism": although the human mind is unpredictable, a poet's heart is much stronger than an 'ordinary' writer's one. A writer could become a terrorist, under certain circumstances, but it would be almost impossible for a (real) poet. It is the poet's heart to make a difference! And I could add that, speaking of our group of poets and a specific kind of poetry that we often nurture, 'Poetry of Witness', it has become an important kind of poetry for most of the poets involved —and this is a good result.

From the same Publisher

(*BE*: Bilingual Editions, English–Italian
Texts translated into Italian by F. Frosini)

Anthologies in Paperbacks:

– 'Fifty-six Female Voices of Contemporary Poetry' – English Ed.;
– 'From an Old Path – Contemporary European Poetry' – English Ed.;
– 'Tunes from the Indian Subcontinent – Contemporary Poetry' – English Ed.;
– 'The Sounds of America – Contemporary American Poetry' – English Ed.;
– 'Whispering to the Heart – Contemporary African Poetry' – English Ed.;
– 'Hues of the World – Contemporary Poetry' – English Ed.;
– 'Singing Together – Poems for Christmas' – English Ed.;
– 'When Love is Bitter' – English Ed.;
– 'United We Stand – Poets Against Terror' – English Ed.;
– 'Homo Homini Lupus: Why To kill a Mockingbird?' – English Ed.;
– 'Time to show up – Poetry for Democracy' – English Ed.;
– 'Through Time, Through Space' – English Ed.;
– 'Let's Laugh Together – Poems for Children' – English Ed.;
– 'Our Only World – Poetry for Planet Earth' - English Ed.;
– 'Winter Melodies' – English Edition.

Anthologies in Ebooks:

– 'At The Crossing Of Seven Winds' – English Ed.;
– 'Nine Tales Of Creation' – English Ed.;
– 'Scattering Dreams & Tales' – English Ed.;
– 'We Are The Words – Siamo Parole' – *BE*;
– 'Whispers to the World – Sussurri al Mondo" – *BE*;
– 'The Double Door' by Daniel J. Brick & Fabrizio Frosini – English Ed.;

– 'Poetry Against Terror'– English Ed.;
– 'How to write Poetry, A Handbook – Come scrivere Poesie, Manuale'– *BE*;
– 'Poets Against Inequality'– English Ed.;
– 'By Land & By Seas – Poetry for the Refugees' – English Ed.;
– 'Voices without veils' – English Ed.;
– 'Singing Together – Poems for Christmas' – English Ed.;
– 'We All Are Persons – Why Gender Discrimination?' – English Ed.;
– 'A Note, a Word, a Brush – Ode to the Arts' – English Ed.;
– 'Seasons of the Fleeting World – Writing Haiku' – English Ed.;
– 'Our Chains, Our Dreams' [Part One] – English Ed.;
– 'Our Chains, Our Dreams' [Part Two] – English Ed.;
– 'Our Chains, Our Dreams' [Part Three] – English Ed.;
– 'Our Only World – Poetry for Planet Earth' – English Ed.;
– 'Time to show up – Poetry for Democracy' – English Ed.;
– 'Let's Laugh Together – Poems for Children' – English Ed.;
– 'Moments of Lightness – Haiku & Tanka' – English Ed.;
– 'United We Stand – Poets Against Terror' – English Ed.;
– 'When Love is Bitter' – English Ed.;
– 'From an Old Path – Contemporary European Poetry' – English Ed.;
– 'Tunes from the Indian Subcontinent – Contemporary Poetry' – English Ed.;
– 'Whispering to the Heart – Contemporary African Poetry' – English Ed.;
– 'Hues of the World – Contemporary Poetry' – English Ed.;
– 'The Sounds of America – Contemporary American Poetry' – English Ed.;
– 'Fifty-six Female Voices – Poetry by Poets Unite Worldwide' – English Ed.;
- 'Homo Homini Lupus: Why To kill a Mockingbird?' – English Ed.;
- 'Essays on the World of Humans' – by D.J. Brick & F. Frosini – English Ed.;
- 'Through Time, Through Space' – English Ed.;
– 'Winter Melodies' – English Edition.

Under publication:
– 'Spring Songs' – English Edition.

Books by Fabrizio Frosini as sole Author:

– «The Chinese Gardens – English Poems» – English Ed. – (published also in Italian Ed.:
  – «I Giardini Cinesi» – Edizione Italiana);
  – «KARUMI – Haiku & Tanka» – Italian Ed.;
  – «Allo Specchio di Me Stesso» ('In the Mirror of Myself') – Italian Ed.;
  – «Il Vento e il Fiume» ('The Wind and the River') – Italian Ed.;
  – «A Chisciotte» ('To Quixote') – Italian Ed.;
  – «Il Puro, l'Impuro – Kosher/Treyf» ('The pure, the Impure – Kosher / Treyf') – Italian Ed.;
  – «Frammenti di Memoria – Carmina et Fragmenta» ('Fragments of Memories') – Italian Ed.;
  – «La Città dei Vivi e dei Morti» ('The City of the Living and the Dead') – Italian Ed.;
  – «Nella luce confusa del crepuscolo» ('In the fuzzy light of the Twilight') – Italian Ed.;
  – «La Chiave dei Sogni» ('The Key to Dreams') – Italian Ed.;
  – «Echi e Rompicapi» ('Puzzles & Echoes') – Italian Ed.;
  – «Ballate e Altre Cadenze» ('Ballads and Other Cadences') – Italian Ed.;
  – «Selected Poems – Επιλεγμένα Ποιήματα – Poesie Scelte» – Greek–English–Italian (Αγγλικά, Ελληνικά, Ιταλικά – Greek translation by Dimitrios Galanis);
  – «Prelude to the Night – English Poems» – English Edition (published also in Italian Ed.:
  – «Preludio alla Notte»).

Under publication:
  – «A Season for Everyone – Haiku & Tanka Poetry» – English Ed.
  – «Il Sentiero della Luna» ('The Moon's Path') – Italian Ed.

Where to find us

Publisher's Page:

https://www.amazon.com/Fabrizio-Frosini/e/B014HA8ZUA/

~*~

Poets Unite Worldwide:

Fb page: https://www.facebook.com/poetsuniteworldwide/

Website: https://poetsuniteworldwide.org

www.ingramcontent.com/pod-product-compliance
Lightning Source LLC
Chambersburg PA
CBHW071513220526
45472CB00003B/1007